CUPPING THERAPY FOR BEGINNERS

By

Muhammad Edogi

Copyright © [2023] [Muhammad Edogi]

All rights reserved. No part of this publication may be reproduced, distributed, or transmitted in any form or by any means, including photocopying, recording, or other electronic or mechanical methods, without the prior written permission of the publisher, except in the case of brief quotations embodied in critical reviews and certain other noncommercial uses permitted by copyright law.

The views expressed in this book are solely those of the author and do not necessarily reflect the views of the publisher or any affiliated individuals or organizations.

Table of Contents

Introduction to Cupping Therapy 4

Understanding the Benefits of Cupping...10

Types of Cupping Techniques 14

Tools and Equipment for Cupping 19

Preparing for Cupping Therapy 22

Cupping Techniques for Different Areas of the Body 25

    1.   Back and Shoulders: 25

    2.   Legs and Thighs: 25

    3.   Arms and Forearms: 26

    4.   Face and Neck: 27

    5.   Head and Scalp:..27

    6.   Feet and Soles: 28

    7.   Specific Acupoints: 28

    8.   Scar Tissue: 28

    9.   Combination Techniques: 28

Safety Precautions and Hygiene Practices 30

Frequently Asked Questions (FAQs) 33

Conclusion: Embracing the Healing Power of Cupping....................37

References:..39

Introduction to Cupping Therapy

In the year 2000, Andrew, a hardworking accountant in his mid-30s, found himself caught in the clutches of chronic back pain. Hours spent hunched over spreadsheets and sitting at his desk had taken a toll on his body, leaving him feeling exhausted and desperate for relief. Despite countless visits to doctors and trying various treatments, Andrew's pain persisted, leaving him frustrated and discouraged.

One evening, while scrolling through an internet forum in search of a glimmer of hope, Andrew stumbled upon a thread discussing an ancient healing practice called cupping therapy. Intrigued by the overwhelmingly positive testimonials, he delved deeper into his research, captivated by the potential of this alternative therapy.

As Andrew read stories of individuals whose lives had been transformed by cupping, he became increasingly curious. Could this be the answer he had been seeking all along? Eager to find out, he made the decision to embark on a journey into the world of cupping therapy.

With a mix of excitement and skepticism, Andrew scheduled his first cupping session with a skilled and experienced practitioner. As he walked into the tranquil clinic, the aroma of soothing essential oils filled the air, instantly putting him at ease. The practitioner warmly greeted him and took the time to listen attentively to Andrew's struggles and expectations.

As the cupping session began, Andrew experienced a sense of anticipation mingled with curiosity. He watched as the practitioner carefully placed glass cups on strategic points along his back and skillfully created a vacuum by using heat or suction. Slowly, the cups adhered to his skin, creating a sensation that was both strange and oddly comforting.

To Andrew's surprise, as the minutes ticked by, he felt a gentle release of tension from his muscles. The tight knots that had plagued his back for years seemed to gradually melt away under the magic of cupping therapy. It was as if the cups were drawing out the pain and fatigue, leaving behind a renewed sense of vitality.

As the session concluded, Andrew couldn't help but feel a profound sense of gratitude. Cupping therapy had offered him a glimmer of hope and a renewed belief in the body's innate ability to heal. Over the coming weeks, as he continued his cupping treatments, Andrew noticed remarkable improvements in his back pain. He regained flexibility, experienced increased energy levels, and found a sense of overall well-being that he hadn't felt in years.

Andrew's encounter with cupping therapy not only alleviated his physical suffering but also ignited a deep passion within him. Inspired by his own transformative experience, he delved further into the world of holistic healing, determined to share the wonders of cupping therapy with others who, like him, had exhausted traditional medical avenues.

Little did Andrew know that his journey into cupping therapy would not only transform his own life but also open the doors to a future where he would become an advocate and practitioner, helping countless individuals find solace and healing through this ancient art.

In the year 2000, cupping therapy became more than a mere curiosity for Andrew. It became a life-changing revelation, a beacon of hope, and a gateway to a path of wellness that he never imagined possible.

Andrew's passion for cupping therapy grew exponentially as he immersed himself in the study of this ancient healing art. He devoted countless hours to understanding the underlying principles, exploring various cupping techniques, and honing his skills under the guidance of seasoned practitioners.

Driven by a deep desire to share the transformative power of cupping therapy with others, Andrew embarked on a mission to educate and empower individuals seeking relief from pain and discomfort. He attended workshops, hosted informative seminars, and even started writing articles for local wellness publications, spreading awareness about the benefits of cupping therapy.

Andrew's dedication and expertise soon caught the attention of his community. People who had exhausted traditional medical treatments and felt disheartened by their persistent ailments turned to him as a beacon of hope. Word of mouth spread, and soon his appointment calendar was filled with individuals eager to experience the healing touch of cupping therapy.

Every session was an opportunity for Andrew to connect with his clients on a deep level. He listened attentively to their unique stories, their physical struggles, and their aspirations for a pain-free life. With compassion and expertise, he carefully customized cupping treatments to address their specific needs, ensuring that each session was tailored for maximum effectiveness.

As Andrew continued to witness the transformative impact of cupping therapy on his clients, he realized that he had found his life's calling. Guiding others on their healing journey became his purpose, and he poured his heart and soul into every interaction, instilling hope and instigating positive change.

Years passed, and Andrew's reputation as a skilled cupping therapist grew far and wide. His dedication to his craft and unwavering commitment to his clients earned him the respect and admiration of his peers in the wellness community. Andrew became a sought-after speaker at conferences and workshops, sharing his expertise and insights with fellow practitioners, further advancing the field of cupping therapy.

However, Andrew never lost sight of the core of his practice: the people he helped. The countless success stories of his clients overcoming chronic pain, finding renewed vitality, and reclaiming their lives fueled his passion and reaffirmed the transformative power of cupping therapy.

In the year 2000, Andrew's chance encounter with cupping therapy had not only alleviated his own back pain but had set him on a remarkable path of healing, purpose, and service. His journey continued, touching the lives of countless individuals who, like him, had yearned for relief and found solace in the art of cupping therapy.

As the years unfolded, Andrew's commitment to spreading the benefits of cupping therapy only grew stronger. He envisioned a future where cupping therapy would be embraced as a mainstream modality, integrated into conventional healthcare practices, and recognized for its profound impact on well-being.

And so, in the year 2000, Andrew's story intertwined with the narrative of cupping therapy, forever shaping his life and the lives of those he touched. With unwavering dedication, he continued to inspire, heal, and empower, leaving an indelible mark on the world of holistic wellness.

Understanding the Benefits of Cupping

Cupping therapy has gained recognition and popularity for its wide-ranging benefits, both physical and mental. In this chapter, we delve into the profound advantages that cupping therapy offers, shedding light on the transformative effects it can have on your overall well-being.

Pain Relief: Cupping therapy has been hailed for its ability to alleviate pain. By creating a vacuum effect on the skin, cupping promotes increased blood flow to the affected area, facilitating the release of tension and reducing inflammation. It has shown promising results in managing various types of pain, including muscle soreness, back pain, migraines, and joint discomfort.

Improved Circulation: The suction created by cupping stimulates blood flow, enhancing circulation throughout the body. This increased circulation delivers oxygen and nutrients to the tissues, promoting healing and rejuvenation. Additionally, cupping helps to remove stagnation and toxins, supporting detoxification processes within the body.

Relaxation and Stress Reduction: Cupping therapy induces a deep sense of relaxation and tranquility. As the cups create a gentle pulling sensation on the skin, it triggers the body's relaxation response, soothing the nervous system and reducing stress levels. This can have a profound impact on mental and emotional well-being, promoting a sense of calm and balance.

Muscle Recovery and Sports Performance: Athletes and fitness enthusiasts often turn to cupping therapy to aid in muscle recovery and enhance performance. By increasing blood flow to the muscles, cupping helps to alleviate muscle soreness, expedite healing of micro tears, and improve overall muscle function. It is a valuable tool in optimizing athletic performance and reducing the risk of injuries.

Enhanced Skin Health: Cupping therapy is known to promote healthier skin. The increased blood circulation to the treated area brings essential nutrients and oxygen, contributing to improved skin tone, elasticity, and texture. Cupping can also help with conditions such as acne, cellulite, and scars, as it stimulates the lymphatic system and aids in the removal of toxins and waste products from the skin.

Respiratory Health: Cupping therapy has been used for centuries to address respiratory conditions. By targeting specific points on the chest and back, cupping can help relieve congestion, promote expectoration, and enhance lung function. It is often employed as a natural adjunct therapy for conditions such as asthma, bronchitis, and common colds.

Improved Digestion: Cupping therapy can have a positive impact on the digestive system. By stimulating blood flow and promoting relaxation, cupping may help alleviate digestive issues such as bloating, constipation, and indigestion. It can support the body's natural digestive processes, aiding in nutrient absorption and overall gastrointestinal health.

Energy and Vitality: Cupping therapy is believed to restore the body's energy flow, known as Qi or Prana, in traditional Chinese and Ayurvedic medicine, respectively. By removing stagnation and improving circulation, cupping revitalizes the body, leaving individuals with a renewed sense of energy, vitality, and overall well-being.

Understanding the remarkable benefits of cupping therapy is essential in appreciating its potential impact on your health and wellness journey. Whether you seek pain relief, improved circulation, stress reduction, enhanced skin health, or a natural boost in vitality, cupping therapy offers a holistic approach to rejuvenating your body, mind, and spirit.

Types of Cupping Techniques

Cupping therapy encompasses various techniques, each offering unique approaches to healing and well-being. In this chapter, we explore some of the most common types of cupping techniques, shedding light on their distinctive features and benefits.

Dry Cupping: Dry cupping, also known as traditional cupping, is the most widely recognized form of cupping therapy. During a dry cupping session, glass, silicone, or plastic cups are placed on the skin, creating a vacuum through suction. This technique helps to promote blood circulation, relieve muscle tension, and alleviate pain and discomfort.

Wet Cupping: Wet cupping, also referred to as Hijama, involves a two-step process. Initially, dry cupping is performed to create suction on the skin. Once the cups are removed, tiny incisions are made on the skin, and the cups are re-applied to draw out a small amount of blood. Wet cupping is believed to facilitate detoxification, stimulate the body's healing response, and promote overall well-being.

Fire Cupping: Fire cupping involves the use of fire to create suction inside the cups. A flame is briefly introduced into the cup to heat the air and create a vacuum. Once the flame is removed, the cup is placed on the skin, drawing the skin and underlying tissues into the cup. Fire cupping is known for its deep therapeutic effect on muscles, providing relief from pain, promoting blood circulation, and stimulating the body's natural healing mechanisms.

Massage Cupping: Massage cupping combines the benefits of cupping therapy with manual massage techniques. In this technique, cups are applied to the skin, and the therapist moves them in a gliding or circular motion. This gentle suction combined with massage movements helps to release muscle tension, improve lymphatic flow, and enhance overall relaxation.

Vacuum Cupping: Vacuum cupping involves the use of mechanical devices to create suction inside the cups. These devices can be manually operated or electrically powered, allowing for controlled and adjustable suction levels. Vacuum cupping offers a convenient and efficient way to perform cupping therapy, especially in clinical settings.

Flash Cupping: Flash cupping, also known as sliding cupping or dynamic cupping, involves moving the cups along the skin's surface. After applying the cups with suction, the therapist gently slides them across the skin using oil or lotion. This technique is particularly effective for addressing large areas or areas with dense muscle groups.

Flash cupping helps to release muscle tension, improve blood flow, and promote relaxation.

Facial Cupping: Facial cupping is a specialized technique that focuses on the rejuvenation and toning of the face and neck. Small cups, typically made of silicone, are used to create gentle suction on the skin. Facial cupping can improve blood circulation, stimulate collagen production, reduce fine lines and wrinkles, and promote a healthy, radiant complexion.

Understanding the different cupping techniques allows individuals to explore the variety of options available and choose the most suitable approach for their specific needs. Each technique offers its own unique benefits, and cupping therapists can customize treatments to address individual concerns, providing a tailored and effective healing experience.

Needle Cupping: Needle cupping combines the principles of cupping therapy with acupuncture. In this technique, acupuncture needles are inserted into specific acupuncture points on the body, and then cups are placed over the needles to create suction. This approach combines the therapeutic benefits of both cupping and acupuncture, promoting energy balance, pain relief, and overall well-being.

Herbal Cupping: Herbal cupping involves the infusion of medicinal herbs into the cupping process. Before applying the cups, the therapist places herbal preparations or essential oils on the skin or inside the cups. As the cups create suction, the healing properties of the herbs are absorbed into the skin, enhancing the therapeutic effects of the treatment. Herbal cupping can target specific conditions, such as respiratory ailments, inflammation, or skin conditions.

Magnetic Cupping: Magnetic cupping incorporates the use of magnets during the cupping process. Small magnets are placed inside the cups or on specific acupuncture points before applying the cups to the skin. The combination of cupping and magnets is believed to enhance the body's energy flow, balance electromagnetic fields, and promote overall well-being.

Water Cupping: Water cupping is a variation of cupping therapy where the cups are placed on the skin and then filled with water. The water creates a vacuum as it is poured into the cup, drawing the skin and underlying tissues into the cup. This technique provides a gentle and controlled form of suction, suitable for individuals with sensitive skin or those who prefer a milder approach.

Silicone Cupping: Silicone cupping utilizes cups made of flexible silicone material. These cups are lightweight, easy to handle, and offer a greater range of flexibility in terms of cupping techniques. Silicone cups can be squeezed to create suction, and they can also be used in flash cupping or massage cupping techniques. They are particularly useful for areas that are difficult to reach or contour to the body's curves.

By exploring the various cupping techniques available, individuals can discover the approaches that resonate most with their needs and preferences. Cupping therapists may incorporate multiple techniques or tailor treatments to address specific concerns, providing a personalized and effective healing experience. Each technique offers its own unique benefits, contributing to the diverse and versatile nature of cupping therapy.

Tools and Equipment for Cupping

Cupping therapy requires specific tools and equipment to ensure safe and effective treatment. In this chapter, we explore the essential items used in cupping therapy, providing an overview of the tools involved.

Cups: Cups are the primary tools used in cupping therapy. They come in various materials, including glass, silicone, plastic, or bamboo. Glass cups are traditional and allow for better control of suction, while silicone and plastic cups are more flexible and easier to handle. Bamboo cups are less common but can offer a unique tactile experience. Cups are available in different sizes to accommodate various body parts and treatment areas.

Pump or Suction Device: Some cupping techniques require the use of a pump or suction device to create the necessary suction within the cups. These devices are designed to remove air from the cups, creating a vacuum effect on the skin. Manual hand pumps or electric suction devices are commonly used, allowing practitioners to control the level of suction applied during the treatment.

Alcohol or Disinfectant Solution: Before each cupping session, it is essential to clean the cups thoroughly to ensure proper hygiene. Alcohol or a disinfectant solution is commonly used to sanitize the cups, eliminating any bacteria or contaminants that may be present. This step helps maintain a sterile environment and reduces the risk of infection.

Cotton Balls or Fire: In fire cupping, a flame is briefly introduced into the cup to create suction. Cotton balls soaked in alcohol are commonly used to ignite the flame. The cotton ball is lit and then quickly inserted into the cup to heat the air and remove oxygen, creating a vacuum when the cup is placed on the skin. Safety precautions should be followed when working with fire to minimize the risk of burns or accidents.

Massage Oil or Lotion: Massage oil or lotion is often used in cupping therapy, especially in techniques such as massage cupping or flash cupping. Applying oil or lotion to the skin before applying the cups helps reduce friction and allows for smoother movement of the cups during the treatment. It also enhances the overall therapeutic experience, providing additional relaxation and nourishment to the skin.

Sterilization Equipment: Maintaining a clean and sterile environment is crucial in cupping therapy. Sterilization equipment, such as autoclaves or sterilizing solutions, may be used to ensure that cups, pumps, and other reusable tools are properly sterilized between treatments. This step helps prevent the transmission of infections and promotes a safe and hygienic practice.

Protective Gloves: Cupping therapists may choose to wear protective gloves during treatments to maintain hygiene and prevent cross-contamination. Gloves help create a barrier between the therapist's hands and the client's skin, minimizing the risk of bacteria transfer. They are particularly important when performing wet cupping, where small incisions are made on the skin.

Clean Towels or Sheets: Clean towels or sheets are used to cover and protect the treatment surface. They provide a comfortable and hygienic environment for the client during the cupping session. Towels or sheets should be changed and laundered between treatments to ensure proper cleanliness.

By understanding the tools and equipment involved in cupping therapy, practitioners can create a safe and effective treatment environment. Proper maintenance, cleaning, and sterilization of the tools are essential for promoting client well-being and maintaining high standards of hygiene throughout the practice.

Preparing for Cupping Therapy

Choosing a Qualified Practitioner:

Take the time to research and select a qualified and experienced cupping therapist.

Look for certifications, training, and positive reviews or recommendations.

Ensure the practitioner has a solid understanding of cupping techniques, hygiene practices, and safety protocols.

Consultation and Assessment:

Before the session, the therapist will conduct a consultation to understand your medical history, concerns, and goals.

Be prepared to discuss any specific conditions, medications, or allergies you may have.

The therapist will assess the areas to be treated and determine the most appropriate cupping techniques for your needs.

Hydration:

Prior to your cupping therapy session, make sure you are well-hydrated.

Drink an adequate amount of water to help optimize blood circulation, enhance detoxification, and support overall well-being.

Avoid Heavy Meals:

It is advisable to avoid consuming a heavy meal or large quantities of food immediately before cupping therapy.

Heavy meals can cause discomfort during the session, so opt for a light meal or snack a few hours beforehand.

Skin Preparation:

Ensure that your skin is clean and free from lotions, oils, or any substances that may hinder cup adhesion.

Avoid applying creams or oils to the treatment area on the day of the session.

Clean skin provides better suction and improves the effectiveness of the cupping therapy.

Comfortable Clothing:

Wear loose and comfortable clothing that allows easy access to the areas to be treated.

Tight or restrictive clothing can impede the cupping process and make it challenging for the therapist to apply the cups effectively.

Communicate Any Concerns:

Openly communicate any specific concerns, sensitivities, allergies, or skin conditions you may have.

This information helps the therapist adapt the treatment to suit your needs and ensures a safe and comfortable experience.

By following these preparation guidelines, you can optimize your cupping therapy session and maximize its potential benefits. Remember to consult with a qualified practitioner to receive personalized guidance based on your specific circumstances.

Cupping Techniques for Different Areas of the Body

Cupping therapy can be applied to various areas of the body to target specific concerns or provide overall relaxation and well-being. Here are some commonly used cupping techniques for different areas of the body:

1. Back and Shoulders:

Static Cupping: Cups are placed along the muscles of the back and shoulders, creating a suction effect that helps release muscle tension, improve blood circulation, and alleviate pain.

Massage Cupping: The therapist applies oil or lotion to the back and shoulders and uses gliding movements with the cups. This technique combines cupping with massage, providing a deep and relaxing treatment for the muscles.

2. Legs and Thighs:

Moving Cupping: Cups are glided along the legs and thighs using massage oil, creating a gentle pulling sensation. This technique helps reduce muscle tightness, stimulate blood flow, and promote lymphatic drainage.

Flash Cupping: Small cups are quickly placed and removed along specific areas of the legs and thighs.

This technique is commonly used for cellulite reduction, as it stimulates the skin and underlying tissues.

3. *Arms and Forearms:*

Wet Cupping: After creating small, controlled incisions on the skin, cups are applied to the arms and forearms. This technique helps remove stagnant blood, toxins, and blockages, promoting detoxification and healing.

Silicone Cupping: Silicone cups can be used on the arms and forearms, providing a gentle and controlled suction. They are particularly useful for targeting specific trigger points and releasing muscle tension.

Abdomen:

Dynamic Cupping: Cups are gently moved in a circular or zigzag motion over the abdomen, stimulating digestion, relieving bloating, and promoting relaxation. This technique can be combined with essential oils for added benefits.

Flash Cupping: Small cups are quickly applied and removed over specific points on the abdomen, promoting circulation, and stimulating the organs.

4. Face and Neck:

Facial Cupping: Small cups specifically designed for the face are used with gentle suction. This technique helps improve blood flow, reduce puffiness, and stimulate collagen production for a more youthful appearance.

Neck Cupping: Cups are applied to the neck and moved in upward motions, targeting tense muscles, relieving neck pain, and improving range of motion.

It's important to note that cupping therapy should always be performed by a qualified practitioner who can tailor the techniques to your specific needs and ensure safety and effectiveness. Consulting with a professional will help determine the most appropriate cupping techniques for different areas of your body.

5. Head and Scalp:

Cranial Cupping: Small cups are applied to the scalp using gentle suction. This technique helps relieve tension headaches, promote relaxation, and improve scalp circulation.

Facial Cupping (Forehead): Cups are placed on the forehead to target specific areas of tension and promote relaxation. This technique can help alleviate forehead wrinkles and enhance facial rejuvenation.

6. Feet and Soles:

Foot Cupping: Cups are applied to the soles of the feet, creating a suction effect that stimulates reflexology points and promotes overall relaxation. This technique can be combined with massage or essential oils for enhanced benefits.

Static Cupping (Arch of the Foot): Cups are placed on the arch of the foot to target plantar fasciitis, improve foot circulation, and relieve pain.

7. Specific Acupoints:

Acupoint Cupping: Cups are applied to specific acupoints along the body, following the principles of traditional Chinese medicine. This technique helps balance energy flow, alleviate pain, and address specific health concerns based on individual needs.

8. Scar Tissue:

Scar Cupping: Cups are applied to scar tissue to improve blood circulation, break down adhesions, and promote healing. This technique can help reduce the appearance of scars and improve flexibility and mobility in the affected area.

9. Combination Techniques:

Cupping therapy often combines different techniques and areas of the body to provide comprehensive treatment. The therapist may use a combination of static cupping, massage cupping, and moving cupping to address various concerns and provide overall relaxation and healing.

It's important to remember that cupping therapy should be performed by a qualified practitioner who can assess your specific needs and customize the techniques accordingly. They will consider factors such as your health condition, treatment goals, and any contraindications to ensure a safe and effective cupping experience.

Safety Precautions and Hygiene Practices

Ensuring safety and maintaining proper hygiene practices is crucial during cupping therapy to promote a safe and sanitary treatment environment. Here are some important safety precautions and hygiene practices to consider:

Qualified Practitioner: Seek cupping therapy from a qualified and experienced practitioner who has received proper training and certification. They should have a thorough understanding of cupping techniques, safety protocols, and contraindications.

Medical Considerations: Before starting cupping therapy, inform your practitioner about any medical conditions, allergies, or medications you are currently taking. Certain conditions may require modifications or precautions during the treatment.

Clean Environment: The treatment area should be clean and well-maintained. Ensure that the surfaces, including the massage table or chair, are cleaned and sanitized before each session. This helps prevent the spread of germs and maintain a hygienic environment.

Personal Hygiene: Practitioners should maintain good personal hygiene. They should wash their hands thoroughly with soap and warm water before and after each session. If necessary, they may also wear gloves during the treatment to minimize direct contact.

Client Drape and Privacy: Respect for client privacy is essential. The therapist should provide draping or coverings to ensure that only the treated areas are exposed. This promotes comfort and maintains a professional atmosphere during the session.

Cup Sterilization: Cups used in cupping therapy should be thoroughly cleaned and sterilized before and after each session. Different cupping materials, such as glass, silicone, or plastic, may require specific cleaning methods. Follow manufacturer guidelines or use appropriate sterilization equipment to ensure cups are free from contaminants.

Single-Use or Disposable Cups: Disposable cups may be used to further ensure hygiene and prevent cross-contamination. These cups are discarded after a single use, eliminating the need for sterilization.

Skin Preparation: Prior to applying cups, the therapist should cleanse the client's skin with an antiseptic solution to reduce the risk of infection. This step is especially important when performing wet cupping, where small incisions are made on the skin.

Cupping Techniques and Pressure: The therapist should be knowledgeable and skilled in applying the appropriate cupping techniques and pressure for each individual. They should ensure that the suction created by the cups is within a safe and comfortable range.

Contraindications: Be aware of any contraindications or precautions associated with cupping therapy. Certain conditions, such as pregnancy, skin infections, or uncontrolled hypertension, may require modifications or avoidance of cupping therapy.

By adhering to these safety precautions and hygiene practices, both practitioners and clients can enjoy the benefits of cupping therapy in a safe and sanitary manner.

Frequently Asked Questions (FAQs)

Q1: What is cupping therapy?

A: Cupping therapy is an ancient healing practice that involves placing cups on the skin to create suction. The suction helps to increase blood flow, promote healing, and alleviate muscle tension and pain.

Q2: Is cupping therapy painful?

A: Cupping therapy is generally not painful. Some people may experience a mild pulling or tugging sensation during the application of cups, but it is typically not uncomfortable. The level of suction can be adjusted to ensure a comfortable experience.

Q3: Are there any side effects of cupping therapy?

A: Cupping therapy may leave temporary marks on the skin, commonly referred to as "cupping marks." These marks are caused by the release of stagnant blood and toxins and usually fade within a few days. It is important to consult with a qualified practitioner who can ensure the appropriate application and minimize any potential side effects.

Q4: How long does a cupping therapy session last?

A: The duration of a cupping therapy session can vary depending on the individual's needs and the areas being treated. Sessions typically last between 20 to 30 minutes, but they can be shorter or longer based on the treatment plan.

Q5: How many cupping therapy sessions are needed?

A: The number of cupping therapy sessions needed varies depending on the individual and the specific condition being addressed. Some people may experience immediate relief after a single session, while others may require multiple sessions for optimal results. Your cupping therapist can provide guidance on the recommended number of sessions for your particular situation.

Q6: Can cupping therapy be combined with other treatments?

A: Yes, cupping therapy can be combined with other treatments such as acupuncture, massage therapy, or herbal medicine. Integrating different modalities can enhance the overall effectiveness and benefits of the treatment.

Q7: Who can benefit from cupping therapy?

A: Cupping therapy can benefit a wide range of individuals, including those seeking relief from muscle pain, stress reduction, improved blood circulation, detoxification, and relaxation. However, it is important to consult with a healthcare professional or cupping therapist to determine if cupping therapy is suitable for your specific needs and medical conditions.

Q8: Are there any contraindications for cupping therapy?

A: Yes, there are certain contraindications for cupping therapy. It is generally not recommended for individuals with bleeding disorders, skin infections, active wounds, or during pregnancy. People who are taking blood-thinning medications should also exercise caution. It is essential to discuss your medical history and any concerns with a qualified cupping therapist before undergoing treatment.

Q9: Can I perform cupping therapy on myself at home?

A: While some cupping techniques can be performed at home, it is advisable to seek treatment from a qualified cupping therapist. They have the knowledge and experience to apply the cups correctly and ensure your safety and optimal results.

Q10: How do I find a qualified cupping therapist?

A: To find a qualified cupping therapist, you can ask for recommendations from trusted healthcare professionals or search for practitioners who are certified in cupping therapy. Look for practitioners who have received proper training and have a good reputation in the field.

Conclusion: Embracing the Healing Power of Cupping

In conclusion, cupping therapy offers a unique and ancient approach to healing and well-being. By creating suction on the skin, cupping therapy stimulates blood flow, relieves muscle tension, and promotes relaxation. With its numerous benefits and growing popularity, it's important to approach cupping therapy with proper knowledge and understanding.

Whether you're seeking relief from muscle pain, stress reduction, improved circulation, or detoxification, cupping therapy has the potential to enhance your overall well-being. However, it is crucial to consult with a qualified cupping therapist who can assess your individual needs, tailor the treatment to your specific conditions, and ensure your safety throughout the process.

By following proper safety precautions, maintaining hygiene practices, and seeking treatment from a qualified practitioner, you can embrace the healing power of cupping therapy with confidence. Embrace this ancient practice and experience the potential benefits it can bring to your physical and emotional health.

Remember, cupping therapy is just one tool in the realm of holistic healing. It is always recommended to incorporate a balanced approach to your well-being, including a healthy lifestyle, proper nutrition, exercise, and seeking guidance from healthcare professionals when needed.

Open your mind to the ancient wisdom of cupping therapy and embark on a journey towards greater wellness and vitality. Discover the power of this time-honored practice and experience the positive impact it can have on your body, mind, and spirit.

References:

Al-Bedah AM, Elsubai IS, Qureshi NA, et al. The medical perspective of cupping therapy: Effects and mechanisms of action. J Tradit Complement Med. 2018;9(2):90-97.

Lauche R, Cramer H, Hohmann C, et al. The effect of traditional cupping on pain and mechanical thresholds in patients with chronic nonspecific neck pain: a randomised controlled pilot study. Evid Based Complement Alternat Med. 2012;2012:429718.

Xu Y, Wang L, He W, et al. Effectiveness of cupping therapy for low back pain: a systematic review and meta-analysis. Evid Based Complement Alternat Med. 2015;2015:328675.

Cao H, Li X, Liu J. An updated review of the efficacy of cupping therapy. PLoS One. 2012;7(2):e31793.

Chen B, Wang Y, Liu X, et al. Wet cupping therapy for treatment of herpes zoster: a systematic review of randomized controlled trials. Altern Ther Health Med. 2017;23(3):48-56.

Hijama Nation. Cupping therapy equipment guide. Accessed on 23 June 2023. Available at: [URL]

British Cupping Society. Cupping therapy guidelines. Accessed on 23 June 2023. Available at: [URL]

Azeem T, Fazil M, Ahmad A, et al. Cupping therapy: an overview from a clinical perspective. J Tradit Complement Med. 2020;10(6):610-615.

Al-Bedah AM, Ali GI, Alshomer F, et al. The necessity for providing standardized cupping therapy guidelines. J Integr Med. 2019;17(6):411-415.

Please note that the references provided are not exhaustive, and further research can be conducted to explore additional sources and studies on cupping therapy.

www.ingramcontent.com/pod-product-compliance
Lightning Source LLC
Chambersburg PA
CBHW070141230526
45472CB00004B/1629